E. Lloyd Jones

Chlorosis - The Special Anaemia of Young Women

Its Causes, Pathology and Treatment

E. Lloyd Jones

Chlorosis - The Special Anaemia of Young Women
Its Causes, Pathology and Treatment

ISBN/EAN: 9783337077310

Printed in Europe, USA, Canada, Australia, Japan

Cover: Foto ©berggeist007 / pixelio.de

More available books at **www.hansebooks.com**

CHLOROSIS:

THE SPECIAL ANÆMIA OF YOUNG WOMEN

Its Causes, Pathology, and Treatment.

BEING

*A REPORT TO THE SCIENTIFIC GRANTS COMMITTEE OF THE
BRITISH MEDICAL ASSOCIATION.*

BY

E. LLOYD JONES, M.D., B.C. (Camb.),

PATHOLOGIST FOR THE STAFF OF ADDENBROOKE'S HOSPITAL, CAMBRIDGE;
LATE RESEARCH SCHOLAR OF THE BRITISH MEDICAL ASSOCIATION,
AND LATE DEMONSTRATOR OF PATHOLOGY IN THE UNIVERSITY
OF CAMBRIDGE.

LONDON:

BAILLIÈRE, TINDALL AND COX

20 & 21, KING WILLIAM STREET, STRAND.

[*PARIS.*   *MADRID.*]

1897.

Towards the expenses of this research grants were made by the Scientific Grants Committee of the British Medical Association in 1893, 1894, and 1895.

PREFACE

THE necessity for making a Report to the Scientific Grants Committee of the British Medical Association is the best excuse which the Author has for publishing this little book.

It does not profess to give anything like a complete account of the disease with which it deals, its object being rather to draw attention to some observations which seem to be new, and to present a theory of chlorosis which shall be consistent with the pathology of the present day.

It is only by the possession of a sound knowledge of the disturbances which affect the healthy blood that we can hope to understand the origin of those kindred disturbances which are met with in disease. For this reason it is my purpose to devote the first part of this report to a consideration of the different conditions of the blood which are coincident with differences of age, sex, temperament, and so on. Having done so, I shall devote the second part to a consideration of some of the characteristic features of chlorosis, seeking to point out the

significance of this special anæmia of young women, and
the relation which it bears to conditions which are merely
physiological.

In the last place, I shall discuss very briefly what prophylactic and therapeutic measures are most likely to
meet with the best success.

My thanks must be expressed to the numerous friends
who have lent their aid in forwarding my work since ten
years ago, when Professor Roy kindly suggested that I
should begin an inquiry into the variations of the specific
gravity of the blood.

CONTENTS

PART I.

THE CHANGES IN THE BLOOD OF HEALTHY MALES AND FEMALES AT DIFFERENT AGES.

PART II.

THE PATHOLOGY OF CHLOROSIS.

PART III.

THE PREVENTION AND TREATMENT OF CHLOROSIS.

CHLOROSIS

PART I.

THE CHANGES IN THE BLOOD OF HEALTHY MALES AND FEMALES AT DIFFERENT AGES.

In former papers* I have described the variations of the specific gravity of the blood in persons of both sexes at all ages, and I showed how these variations were related to differences in the number of corpuscles and in the amount of hæmoglobin. In the papers referred to I showed that the mean specific gravity is 1050 in boys of two to three years old, and rises to 1058 at seventeen years of age. It remains 1058 during middle life, but falls in old men. In females the mean specific gravity is found to be the same as that in the male until after puberty; after the seventeenth year it falls, while that in the male is still rising, remaining low until twenty-five years of age, when it rises to 1055 or 1056. About the climacteric the mean specific gravity is found to attain its maximum in the female (1057).

Charts I. and II. are similar to those published by me

* *Journ. Physiol.*, 1887 and 1891.

in the *Journal of Physiology*,* but with the addition of new observations, and they show the results of measurements of the specific gravity of the whole blood, made on about 1,400 healthy individuals.

Chart I. gives the results of observations made on male, Chart II. those made on female subjects. In each chart an observation is recorded by a black dot; the abscissæ denoting ages, the ordinates the specific gravity of the blood.

The most striking difference between Charts I. and II. is perhaps the remarkable broadening out of the band of dots in females about the age of puberty. This is chiefly due to the frequent occurrence of a low specific gravity of blood about this period of life. Some of the persons from whom the observations recorded in Chart II. were taken were, I believe, really chlorotic, although they did not complain of ill-health. Eliminating these cases, we still find that in the case of the majority of young women the specific gravity of the blood falls at about this age.

In the paper to which I have referred these variations of the specific gravity of the blood were shown to be closely related to variations in the amount of hæmoglobin and of the number of corpuscles. I have commenced a series of observations on the percentage of hæmoglobin in the blood of healthy persons of all ages—not that this field of inquiry has been neglected, but because it seems to be of importance that a much larger number of experiments should be made.

I have also been measuring the relative proportions of corpuscles to serum in the blood of healthy persons. The method employed consists in whirling a tube filled with the blood at a speed of 10,000 revolutions a minute. So far as I am aware, Dr. Haviland first applied this centrifugal method to the examination of drops of blood.

The results of observations on fifty persons past the age of puberty, and not over twenty-six years of age, gave the following averages:

* *Journ. Physiol.*, vol. xii., p. 299, 1891.

Proportion of Length of Column of Corpuscles to Length of Column of Serum.

In males ... ... :: I : I˙0I
In females ... ... :: I : I˙22

These figures shew that the average ratio between the length of the column of serum and that of the column of corpuscles is greater in females after puberty than in males of the same age, although the difference is not great.

This difference does not exist before the age of puberty.

Observations on the specific gravity of the serum obtained by this method are not difficult to make, and they shew that the average specific gravity of the serum is rather higher in the female after puberty than in the young girl, and rather higher in the female than in the male.

The Changes in the Blood of Healthy Females at the Age of Puberty.

From the charts I present it will be seen that the blood of the female undergoes a marked change after puberty, the specific gravity becoming lower than in the male, and the mean specific gravity being also lower after the age of sixteen years than it is during girlhood ; and the results to which I have already referred shew that this change is due to a relative diminution of the number of corpuscles, and of the amount of hæmoglobin, the amount of plasma being relatively larger than before the age of puberty.

This change in the blood of the female occurs just after the age when menstruation sets in, and one is led to ask whether menstruation may account for it.

On making observations in order to determine the influence of menstruation in females, I found that, while in some persons menstruation appeared to have no effect

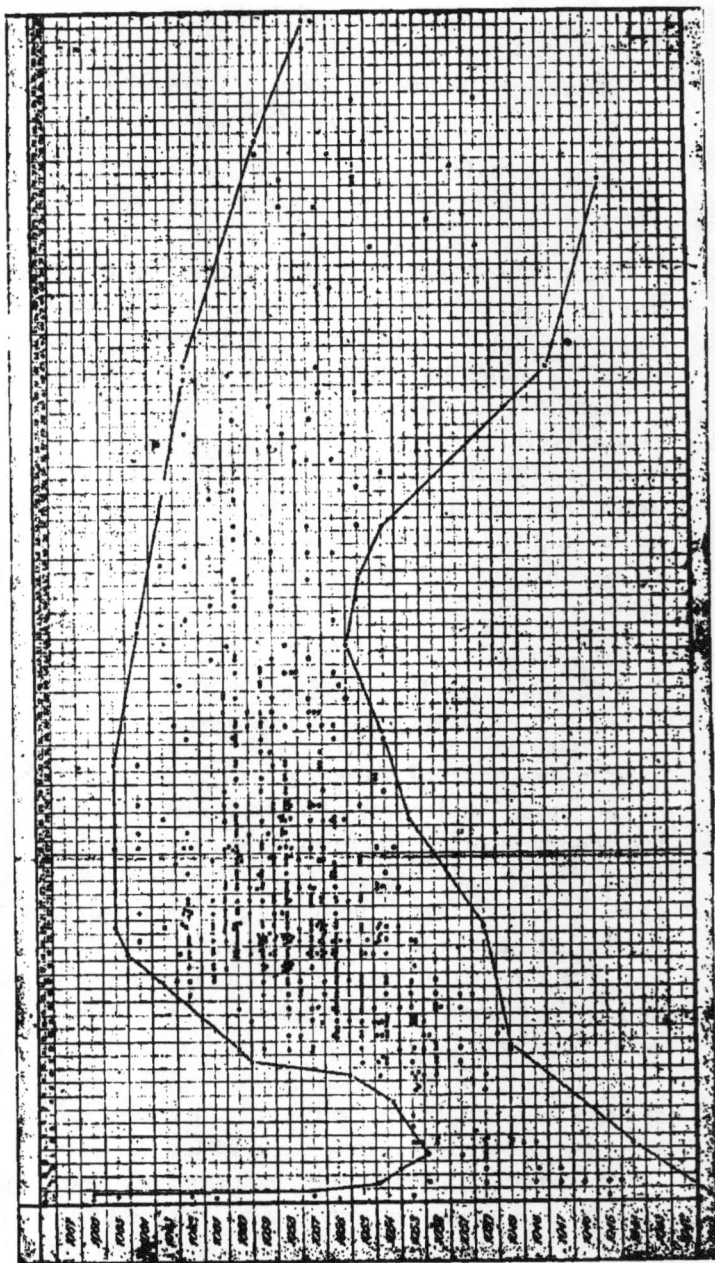

CHART I., SHOWING THE RESULTS OF OBSERVATIONS ON THE SPECIFIC GRAVITY OF THE BLOOD OF HEALTHY MALES
OF DIFFERENT AGES AND THE UPPER AND LOWER LIMITS OF VARIATION CONSISTENT WITH HEALTH.

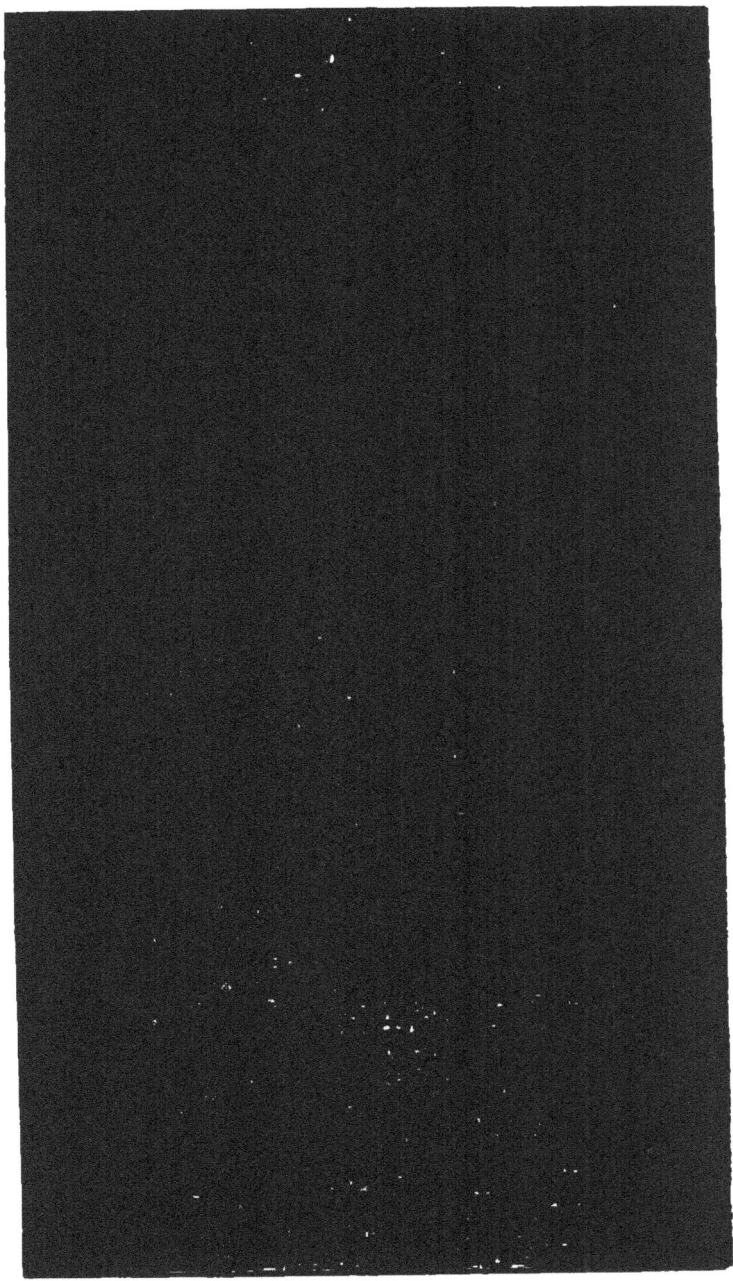

CHART II., SHOWING THE RESULTS OF OBSERVATIONS ON THE SPECIFIC GRAVITY OF THE BLOOD OF HEALTHY FEMALES OF DIFFERENT AGES AND THE UPPER AND LOWER LIMITS OF VARIATION CONSISTENT WITH HEALTH.

on the specific gravity of the blood, in other cases the specific gravity of the blood became less.

Different observers (Schmaltz,[*] Hayem,[†] Reinert,[‡] Bohnstedt,[§] Vierordt,[||] and others) have found that the amount of hæmoglobin is diminished, remains unaltered, or even becomes increased during menstruation; but the matter has not received the attention it deserves; no one, so far as I am aware, has made a sufficient number of observations, though this question might be finally settled by making a series of careful observations on many different subjects.

Schrader[¶] made some careful observations which showed that about or at the menstrual period less nitrogen was eliminated both by the urine and the faeces, the diet being duly regulated; but this change did not always set in at the same time; sometimes it occurred during the whole of the menstrual period, sometimes at the commencement, and sometimes before the appearance of the catamenia. It would be interesting to know if these changes correspond with changes in the blood.

Coincident with the changes which take place in the blood of the female at puberty, and probably as the result of these, tissue change is lessened[**]; for the amounts of CO_2 and of urea given off, compared with the body weight,

---

[*] 'Untersuchung des sp. gew. des menschlichen Blutes,' *Leutsch. Archiv. f. Klin. Med.*, Bd. xlvii., s. 153.

[†] *Du Sang*, Paris. 1889, p. 193.

[‡] *Die Zählung der Blutkörperchen*, Leipzig, 1891.

[§] 'Zählungen der roten Blutkörperchen bei verschiedenen pathologischen Zuständen,' Inaug. Diss., Breslau, 1889, s. 8.

[||] Vierordt, 'Leitrag. z. Physiol des Blutes,' *Arch. f. Physiol. Heilk.*, 1854, Heft. 2, s. 259.

[¶] Schrader, 'Untersuchungen über den Stoffwechsel während der Menstruation,' *Zeits. f. Klin. Med.*, Bd. xxv., s. 72.

[**] Landois and Stirling's *Text Book of Physiology*, third edition, 1888, pp. 188, 189.

are less in the woman than in the man, while the incidence of puberty in the woman is associated with a marked tendency to store up fat.

In other words, in the woman katabolism is less active than in the male, in order, probably, that she may reserve material to serve as a store for the fœtus in utero in the event of pregnancy. A similar kind of blood-change is repeated in some healthy women at each menstrual period, making it appear likely that at such times something happens which leads to fresh additions to the reserve laid aside.

Influence of Constitution, and Relations between Temperament, Blood-composition, and Fertility.

At an early stage of my work I found that there are very considerable differences in the blood of different healthy individuals of the same age and sex. These are shown by differences in the specific gravity, in the amount of hæmoglobin, in the number of corpuscles, or the proportion of corpuscles to serum, and in the occurrence of a greater daily variation of the specific gravity in some persons.

The blood in any healthy person, although always undergoing changes, possesses a fairly constant mean composition, and upon this constancy depends the value of observations such as those offered in this paper.

Since, according to my own showing, the blood is subject to daily variations, I am sometimes asked whether conclusions from a single examination are substantially useful for purposes of statistics. This criticism demands a careful reply. In some persons, it is true, the daily variations range so widely that one may form an erroneous impression from a single examination of the blood ; in others, the daily variations lie within very narrow limits. Indeed, the variations in the majority of persons are within such narrow limits that a single examination of the blood gives a result which is sufficient for such purposes as those I have in view, and a repetition of the observation is, as a rule, needless.

Formerly, I used to repeat my observations on each individual, and I used to note the hour when the observation was made; but little advantage was gained by doing so. Now I do not, unless the blood be taken before breakfast, or there be some special circumstance rendering such a note necessary.

The amount of daily variation differs in different individuals. In certain kinds of persons the specific gravity of the blood oscillates within very wide limits. In a previous paper I gave portions of the daily curves taken from two individuals selected as extreme instances; the one was a good instance of the person who had a changeable specific gravity of the blood, and the other of a person whose blood was constant in specific gravity.

I can only account for this difference by supposing that the total volume of blood is greater in those persons whose blood specific gravity changes little from day to day. If the volume of blood in the circulation is great, it necessarily follows that less alteration in its composition will follow those incidents which are continually tending to alter it. For example, the drinking of water will clearly produce less effect on the specific gravity of the whole blood if the volume of blood be great. Again, profuse perspiration will raise the specific gravity of the blood through more degrees the less the antecedent volume of the blood.

For the present I forbear to deal any further with the variations of the blood-composition in individuals. A large series of observations has shown me that, with average persons, a single observation may be utilized in place of a series of observations, and that the changes in each individual are within such narrow limits that they may, for our purposes, be neglected. But at the same time, the fact must not be ignored that a single observation does not always give us a true estimate of the average state of the blood in an individual.

Disregarding, then, the variations of the blood in the same individual, the mean specific gravity in different individuals varies in a remarkable way. I have already shown in a previous paper that in man the specific gravity of the blood is lower in some persons than in others ;* so that, what may be an abnormal specific gravity in one, becomes a normal specific gravity when observed in another kind of person. I pointed out that social position and conditions of life influence the composition of the blood, and I showed that those who had light eyes, hair, and complexion have generally a lower blood specific gravity than those who are dark.

During the last eighteen months I have been able to collect evidence which shows that the composition of the blood has a curious relation to fertility.

Examining the blood of healthy individuals, I noticed that those who had a high specific gravity of the blood had generally fewer brothers and sisters than those who had a low specific gravity of blood. I have examined the blood of males and females with the object of ascertaining how far this rule holds good.

To put the matter to a further test, I made observations on one hundred males whose parents both survived until many years after the birth of the last child—a time when they may be supposed to have exhausted their fertility— and found that those whose blood had the lowest specific gravity had an extraordinarily large number of brothers and sisters ; while those whose blood had a high specific gravity had fewer brothers and sisters.

The average number of brothers and sisters was as given below :

* *Journ. Physiol., loc. cit.*

Specific Gravity of Blood.			Number of Brothers and Sisters.
1056-1056·9	...	...	8-9
1057-1057·9		...	7-8
1058-1058·9	...	...	6-7
1059-1059·9	...	...	5-6
1060-1060·9	...	...	5-6
1061	...	...	4-5

Here it may be remarked that in women the specific gravity of the blood is lower in the years immediately succeeding puberty, but children begotten in early life have not a lower specific gravity of blood than those begotten in later life.

The observations given below were all made during the past year on single women between the ages of eighteen and twenty-seven; but it is difficult to collect many observations, because one has to avoid those married women who have borne children, and those who are pregnant or suckling.

The average number of brothers and sisters was as given below:

Specific Gravity of Blood.			Number of Brothers and Sisters.
1052-1055·9	...	...	7-8
1056-1057·9	...	...	6-7
1058	...	...	5-6

It appears that in women and in men a low specific gravity of the blood is found in those whose mothers were prolific; so that, by examining the blood of a man or woman, one may forecast the number of brothers and sisters. Those women who have the lowest specific gravity of the blood, and the largest number of brothers and sisters, are generally fair, with pretty pink and white complexions, blue or light eyes, fine hair, with a goodly amount of subcutaneous fat.

Such women are, I believe, unusually prolific, and they and their sisters are liable to suffer from chlorosis.

At present I am still engaged in collecting observations on the blood in families, in order to ascertain the manner in which the blood characters pass down from father or mother to children.

It appears to me to be of great importance that a certain kind of blood should be associated with fertility, and another kind with comparative infertility ; the fact suggests some reflection as to the causes of sterility. At the same time, these observations recall the fact which I showed in my Report last year, namely, that the chlorotic' women have many brothers and sisters.

So far as we have considered the condition of the blood in the two sexes, we have seen that at the age of puberty the blood in the two sexes is differentiated, the specific gravity, the number of corpuscles, and the hæmoglobin become diminished, and the amount of plasma undergoes a relative increase. We have seen that this change occurs about the time when menstruation begins, and we have seen that a slight change of the same character is noticeable in many women at each menstrual period. We have seen that among healthy persons some possess blood which exhibits in an unusual degree some of the characteristics of the blood of the woman, and these persons, whether male or female, come of prolific mothers. Bearing these facts in mind, we shall now consider chlorosis, a disease which is almost, if not absolutely, confined to women, and is evidenced by a blood-change occurring at the same time of their life as the blood-change of which we have been speaking.

PART II.

THE PATHOLOGY OF CHLOROSIS.

The Blood-changes in Chlorosis.

THE blood-changes form such a prominent feature in cases of chlorosis that they may justly be given the first place in a description of some of the clinical features of the disease. Those most characteristic of chlorosis are the following, firstly, *diminution of the specific gravity of the blood*. This is well shown in Chart III., which is constructed on the same plan as one published in 1891,* and in which each observation on a chlorotic person is represented by a dot. On the same chart lines are drawn showing the upper and lower limits of variation of the specific gravity in healthy young women, and another line gives the mean specific gravity of the blood in healthy young women at different ages.

It will be noticed that the specific gravity in each case of chlorosis was below the mean specific gravity of healthy women, and the recording dots form a prolongation downwards of the dip which the lower limit of variation exhibits in women between the ages of fourteen and about twenty-six years.

It thus appears that the blood of the chlorotic subject exhibits an exaggeration of the fall in specific gravity which occurs in healthy young women about the age of puberty; and, indeed, it is likely that I included among my healthy women some who were in reality chlorotic.

* *Journ. Physiol.*, vol. xii., 1891.

CHART III.

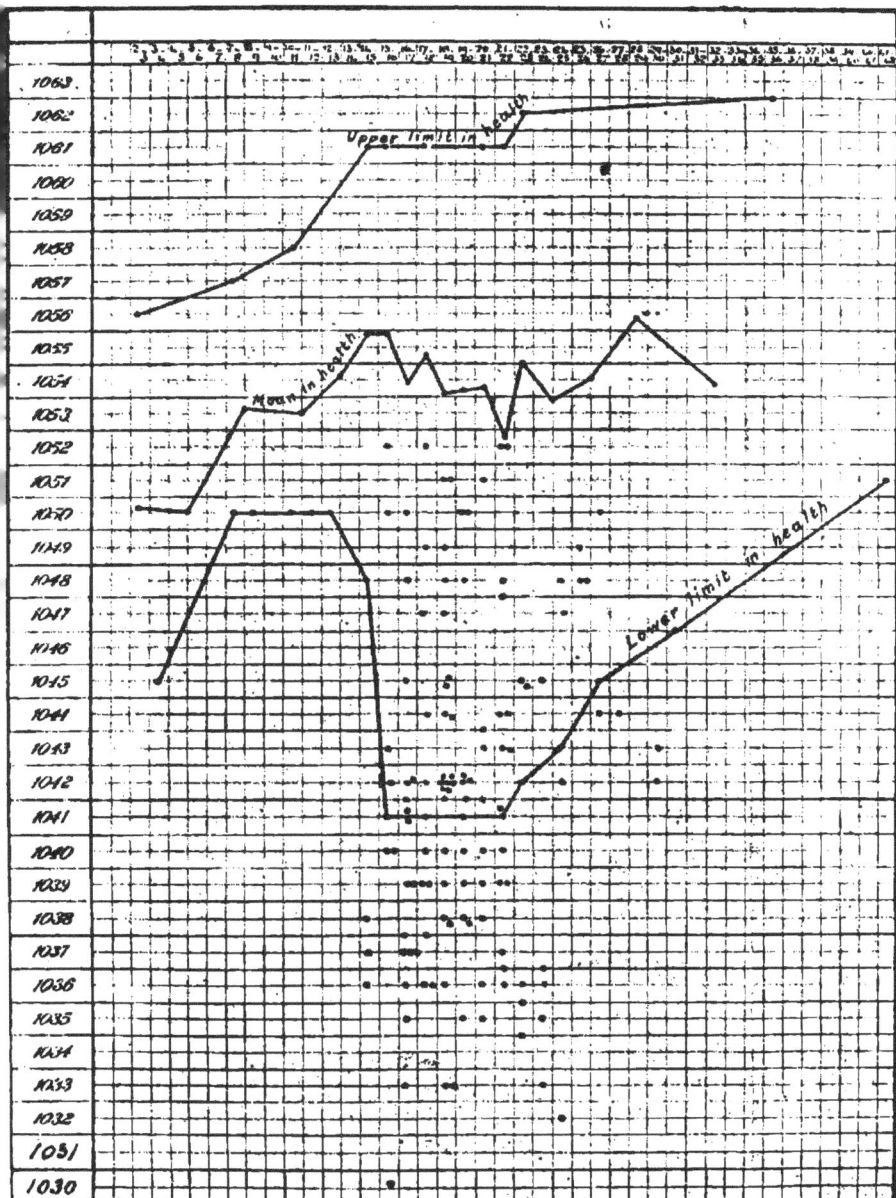

SHOWING THE VARIATIONS IN THE BLOOD SPECIFIC GRAVITY IN HEALTHY
FEMALES FROM TWO TO FORTY-TWO YEARS OF AGE, AND THE RESULTS
OF OBSERVATIONS UPON 120 YOUNG WOMEN WITH CHLOROSIS.

2—2

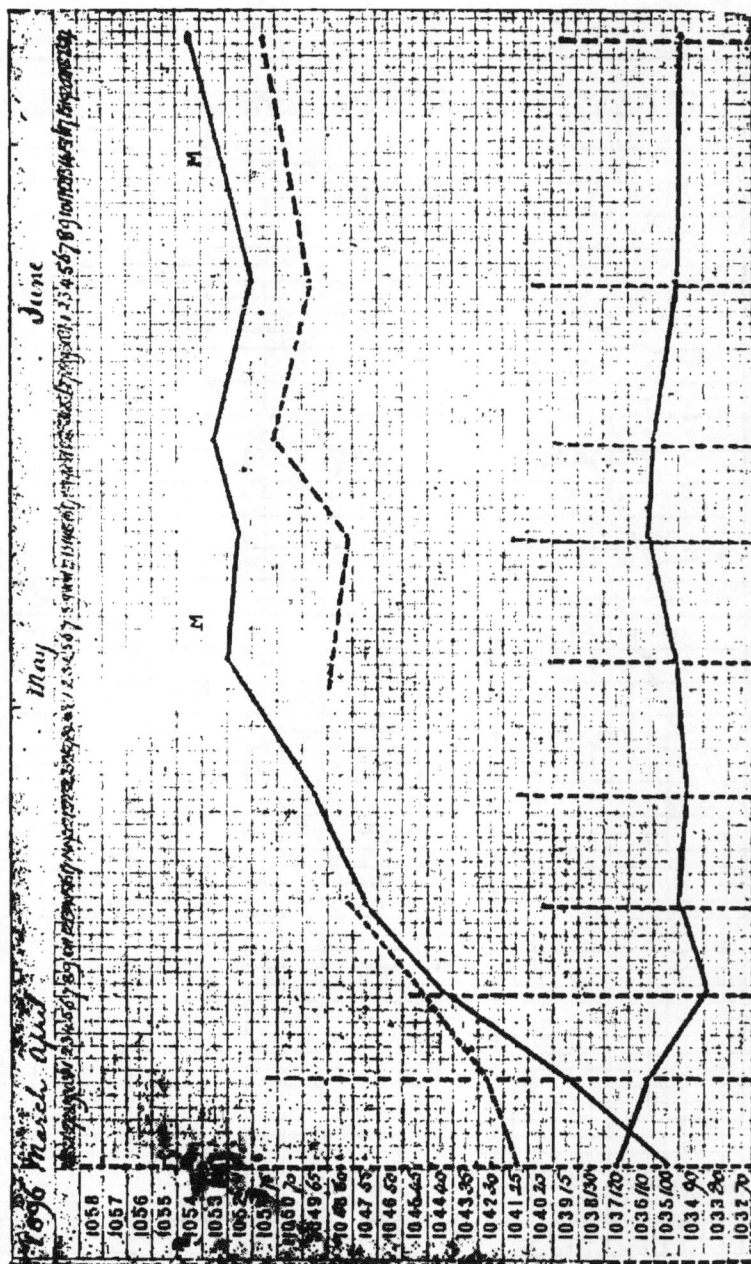

CHART IV.

Equivalent of metallic iron, two grains.　　Six bipalatinous daily.

Hæmoglobin in Chlorosis.

Secondly, the amount of hæmoglobin is always reduced in cases of chlorosis. It may be reduced so low as 17 per cent. on Fleischl's scale.

In twenty-one severe cases the percentages of hæmoglobin before treatment were: 17, 22, 24, 25, 27, 30, 31, 32, 33, 35, 35, 35, 38, 43, 44, 44, 45, 48, 48, 50, and 58, respectively.

Number of Red Corpuscles.

Thirdly, diminution of the number of red blood corpuscles. Hayem[*] states that chlorosis is attended by (1) diminution of the number of red corpuscles; (2) a deficiency of hæmoglobin in each corpuscle; but he recognises two kinds of chlorotic blood. The commoner cases, according to him (about 90 per cent.), show alteration in the individual corpuscles, and the remaining ones (about 10 per cent.) show slight changes in the individual corpuscles, but diminution of the number of these. He states that as new corpuscles appear during the process of recovery, the second condition passes into the first.

Among my cases a marked reduction of the *volume* of corpuscles was exceedingly commonly observed (see Charts IV., VI., and VII.). So great is this reduction in the volume of corpuscles that it cannot be explained by the presence of microcytes, but must be attributed to an absolute reduction in the number of these.

Proportion of Serum to Corpuscles.

The enumeration of the corpuscles involves the expenditure of so long a time that I have been quite unable to make weekly observations on my chlorotic patients with the hæmacytometer; but it is quite possible to measure the volume of corpuscles to serum with something like an approach to accuracy by the method described in the first part of this paper, and this I have systematically done. I

[*] *Du Sang*, Paris, 1889, p. 617, *et*

always express the result in figures which indicate the length of the column of serum when a column of corpuscles an inch long is regarded as the unit.

The proportion of lengths of the columns in healthy young women are as 1 : 1·2, as shown in the first part of this paper. This proportion was always exceeded in the following twenty-six cases of chlorosis:

Case.	Proportion of serum to Corpuscles.	Percentage of Hæmoglobin.	Specific Gravity.
F. B. ...	1·28 ...	... —	1040
E. B. ...	: 1·46 ...	... 58	1049
T.	: 1·5 ...	Obstinate case ...	1049
E. T. ...	1·6 ...	... 35 ...	·1037
E. N. ...	1·6 ...	... 48	1045·2
N. L. ...	1·66 ...	Obstinate case ...	1048
F. C. ...	1·8 ...	... 43	1040
F. G. ...	2·0 ...	... 44	1041·5
J. A. W. ...	2·0 ...	... — ...	1041
K. C. ...	2·0 ...	... —	1044
A. C. ...	: 2·2 ...	... 48	1043
H. A. ...	: 2·2 ...	45	1040
S. B. W. ...	: 2·3 ...	... 44	1045
A. L. ...	: 2·37 ...	... 30	1037
S.	2·39 ...	Obstinate case ...	1048
L. B. ...	: 2·48 ...	... 27	1036
Mrs. B. ...	: 2·49 ...	... 38	1038
T.	2·6 ...	... —	1042
A. R. ...	: 2·78 ...	... 25 ... :..	1036·5
Mrs. G. ...	2·94 ...	Obstinate case ...	1042
A. T. ...	: 3·0 ...	35 ; Obstinate case ...	1036
E. C. ...	3·21 ...	... —	1042
M. M. ...	: 3·79 ...	... 35	1038
M. S. ...	3·8 ...	... —	1038
F. S. ...	4·25 ...	... —	1042
S. K. T. ...	5·0 ...	... 25	1035
N. R. ...	5·2 ...	... 17	1030

These figures give 2·49 as the average proportion between the length of the columns of corpuscles and that of the serum in cases of chlorosis. While this considerably exceeds the proportion in healthy young women, in a few of these cases the proportion differed little from the normal. Further, while I never found a healthy young woman with a proportion of serum so great as this—the highest proportion among healthy young women being under two—yet in some few of the above cases of chlorosis, e.g., E. N., N. L., E. B., T., F. B., the proportion was within the limits observed in health. In all these cases of chlorosis, excepting seven of them, the proportion of serum to corpuscles was increased, signifying most probably an absolute reduction of the number of corpuscles. As the patients recovered the proportion of serum to corpuscles always became normal.

These observations distinctly show *that the reduction of the number of red corpuscles in many of these cases of chlorosis was considerable*, and by reference to Charts IV., VI., VII. it will be seen that improvement was indicated as much by an increase of the number of corpuscles as by an increase of the hæmoglobin and of the specific gravity of the whole blood. There is nothing in my experience more characteristic of chlorotic blood than this *extreme* diminution of the volume of red blood corpuscles, together with an equally-marked reduction in the amount of hæmoglobin.

The specific gravity of the serum obtained from chlorotic blood, as a rule, differs little, if at all, from the serum of the healthy, while it may be increased or more rarely diminished. This condition of the serum in chlorotics shows that the marked diminution of the specific gravity of the whole blood in chlorosis must be largely due to either an absolute diminution of the number of red corpuscles, or an absolute increase of the amount of plasma.

I have examined the blood microscopically in some

of my cases of chlorosis ; but I do not think that my experience is sufficiently great to warrant my publishing the results, for through inexperience one is often deceived with regard to the microscopic appearances of blood. Hayem has dealt with this part of the subject in his excellent account of chlorosis,* and he lays special stress on the occurrence of smaller red corpuscles while recovery is taking place.

The Total Volume of Blood.

It would be of much value if we could estimate the total volume of blood in cases of chlorosis.   I have noticed that chlorotic girls generally bleed easily and copiously, but that the contrary is often observed in chronic cases, such as those of E. B. and Mrs. G. (Charts XI. and XII.).

Rubenstein,† in a paper dealing with the salutary effects of the withdrawal of blood in cases of chlorosis, states that in his opinion the quantity of blood in circulation is increased.   I am inclined to believe that this is the case in many cases of chlorosis, and such a view is consistent with my own views as to the significance of the disease.   It has been maintained by some that there is in healthy women a tendency to the storing up of blood during the intermenstrual periods.   If such be the case, it seems not unlikely that chlorosis may be an expression of an exaggeration of this process.

The Relations between Chlorosis and other kinds of Anæmia.

It appears, therefore, that chlorosis is a disease in which the blood undergoes changes of a definite kind.   What

---

* Hayem, *Du Sang*, 1889, p. 621.

† ' Ueber die Ursache der Heilwirkung des Aderlasses bei Chlorose,' *Wien. Med. Presse*, 1893, Nos. 33 and 34.

these are in a typical case of chlorosis may be seen from the charts and from the facts already given. In examining a case of anæmia, I attach great importance to the following observations :

1. The percentage of hæmoglobin.

2. The number of corpuscles per cubic millimetre, or the proportion of corpuscles to serum after centrifugalizing.

3. Specific gravity.

4. The colour of the lips, ears, and complexion.

5. The readiness with which the patient yields blood from a puncture on the finger.

In ordinary chlorosis the percentage of hæmoglobin, the proportion of corpuscles to serum, and the specific gravity of the blood are reduced usually to a great extent, and this reduction is so marked as to constitute an important element in diagnosis.

Chlorosis and Oligæmia.

In all cases of simple chlorosis the lips and ears are pale, and as a rule the patient yields blood very readily from a puncture on the finger. This condition of blood hardly ever or never occurs in young men unless they suffer from some serious organic disease. Young men may suffer from pallor, but the blood of a pale young man and of a chlorotic girl, as a rule, differ greatly. In a young man who suffers from marked pallor, but is free from organic disease, we oftenest find (contrary to what we find in chlorosis) that the percentage of hæmoglobin is actually increased. It is often greatly increased, and the proportion of corpuscles to plasma is increased, and though the complexion be pale, the lips and ears are red. Moreover, the patient does not yield blood readily from a puncture on the finger. This condition (deficiency of plasma) is the commoner fault in young men, while chlorosis is the commoner in young women ; but deficiency of plasma, with or without chlorosis, may occur in young women.

I wish to emphasize my belief that chlorosis and the deficiency of plasma of young men and young women are examples of two widely different kinds of anæmia. In chlorosis I believe that the quantity of plasma in the body is often increased absolutely, as well as relatively; but in the other cases of anæmia, which occur in persons of either sex, since they bleed less readily from a puncture on the finger, and for certain other reasons, I am inclined to believe that there is also an *absolute* diminution of the amount of plasma in circulation.

I believe that this condition may reasonably be termed 'oligæmia.' When oligæmia, in this sense, co-exists with or supervenes in the course of chlorosis, the signs of chlorosis become much obscured. In those oligæmic persons who are pallid, but whose lips and ears are red, the number of corpuscles is relatively increased, and the specific gravity of the blood is increased, the amount of blood-plasma being less than normal. In another paper* I stated that for such patients rest, dieting, and change of air were the best and the only potent therapeutic measures which I could recommend, and I pointed out the uselessness of treating such cases with iron.

Of late I find that when such persons, whose blood specific gravity is generally too high, take chloride of sodium, the amount of plasma becomes increased, and the specific gravity falls. The fall in specific gravity apparently is not due to an actual diminution of the number of red corpuscles, or to a diminution of the amount of the hæmoglobin, but to an increase in the total quantity of plasma. I do not think that the salts of potassium have this effect; in fact, I am inclined to believe that as a rule the salts of potassium have the effect of lessening

* 'The Diagnosis and Treatment of Headaches Accompanied by Diminished or Increased Blood-Pressure,' *Practitioner*, 1889.

the volume of plasma; but I hope to test this point fully. At present I can recommend the administration of sodium chloride and strychnine as likely to be of use in cases of oligæmia of this kind. I find that this condition of blood occurs frequently, if not constantly, in chronic epileptics; and in such cases I think that bromide of sodium should be used rather than the potassium salt.

Young men or young women who suffer from oligæmia have generally none or few brothers and sisters. Chlorotic young women have generally many brothers and sisters. *These two kinds of anæmia bear an important relation to the two kinds of blood which occur in healthy persons to which I have already referred. Chlorosis is an exaggeration of the one condition, oligæmia*—in my sense of the term—*of another.*

Chloro-Oligæmia.

Some cases of chlorosis are, as I have already stated, complicated with the coexistence of oligæmia—that is, with a diminution of the quantity of plasma. These persons have usually, but not invariably, been chlorotic for a considerable time; they have gone from bad to worse; they have generally wasted and lost the typical chlorotic aspect, and their blood no longer shows the signs of severe or acute chlorosis, for these signs are masked by the diminution in the amount of plasma. If a chlorotic patient who has only 30 per cent. of hæmoglobin begins to lack blood-plasma, the percentage of hæmoglobin will increase, the specific gravity will increase, the relative volume of corpuscles to serum will increase, but the patient gets worse. She will still look pale, her ears and lips, however, will have a deeper red colour than while she suffered from chlorosis pure and simple, but she will become more liable to temporary pallor. Such a patient has chloro-oligæmia : she has something more than ordinary chlorosis.

The signs of chlorosis are masked or disfigured when

such changes supervene, and this, I take it, is why different medical men apply the word 'chlorosis' to cases which differ markedly.

Some other Signs of Chlorosis.

It would be quite impossible to give anything like a full account of the other signs of chlorosis in a paper of but moderate length, and those only will be referred to which have engaged my attention, or which appear to me to throw most light upon the etiology of the disease.

First of all, the aspect of the patient in an uncomplicated case of chlorosis is very characteristic. The face is often yellow or pale, but the complexion may be pink and white. Many chlorotic women have a very pretty pink and white complexion.

Lawson Tait speaks of chlorosis as the 'anæmia of good-looking girls.' Dr. Frederick Taylor* has observed a predilection for wearing green ribbons, and an aversion to pink. In my experience the chlorotic girl, when thoroughly cured, has a remarkably pretty complexion, which is better suited by green than any other colour. I believe that a girl thoroughly cured of simple chlorosis is always pretty—remarkably pretty. If my views regarding the significance of chlorosis and its relation to fertility are correct, it is intelligible why this complexion should be regarded as pretty.

In almost all cases the lips and ears are pallid; and, indeed, while pallor of the cheeks is but little guide, pallor of the lips and ears is a very valuable guide in forming an opinion as to the amount of hæmoglobin in the blood.

The waxy pallor of the typical chlorotic is little liable to sudden aggravation. I believe that the vessels of these patients are well filled, and that therefore they do not get pale from emotional and other causes, as chloro-oligæmics

* 'A Discussion on Anæmia,' Brit. Med. Journ., September 19, 1896, p. 719.

and oligæmics do. The flushing of their faces gives them the pretty pink and white colour. The chloro-oligæmic girl, or the oligæmic young man or woman, does not blush readily, but is liable to sudden attacks of marked pallor; the blood is perhaps insufficient to fill the vessels under certain conditions, and in these persons one may often observe capillary pulsation.

A pale face with pale lips, blue or white sclerotics and blanched ears, always indicates a low percentage of hæmoglobin and a relative excess of plasma, or a deficient number of red blood corpuscles. In chlorosis this kind of pallor exists, but the pallor may be very slight, and may only exhibit itself as a want of the normal depth of tinting of the lips or ears. In some of these cases, if the patient's face be flushed, the anæmia may easily pass unrecognised if the blood be not examined; but the want of depth of colouring is always to be recognised with practice, and an examination of the blood at once gives evidence of the existence of anæmia. I find that young children very often exhibit an anæmia of this kind, in which the blood resembles that of chlorotic persons, while it yields to the same treatment.

The pupils in chlorotics are often dilated, so that the eyes look bright. This is most noticeable in those whose faces are flushed. There may be some puffiness due to œdema, or an appearance of puffiness due to abundance of subcutaneous fat. The eyes are often light (according to my experience), but by no means invariably so. Wunderlich has noticed that chlorosis is more frequent in blondes than brunettes, and that it occurs in strong and well nourished as well as in weakly persons. The hair is fine, the hands often show pigmentation at the roots of the nail and over the joints, and there is often a fine tremor of the extended fingers. The neck is full, and the thyroid is often enlarged visibly; it is nearly always found to be so on palpation.

I shall make no reference to the state of the heart

beyond this : that there is nearly always some increase of the cardiac dulness, as pointed out by Stark, and that the pulse is constantly accelerated.

The pulse-rate and the specific gravity of the blood in twenty-nine cases of typical chlorosis are given in subjoined table :

Initial.	Pulse-rate.	Sp. gr. Blood.
A. M.	142	1037
F.	139	(?)
C.	134	1039
H.	132	1040
R.	132	1036
R.2	132	1030
K.	128	1044
M.	126	1038
F.	124	1039
A.	120	1036
E.	120	1049
A.	116	1043
A.	113	1036·5
E.	108 ...	1036·2
G.	108	1039
S.	106	1035
L.	105	1041
L.	104	1036
C.	104	1047·5
E.	104	1043
F. G.	100	1041·5
F.	100 ...	1042
E.	100	1037
Mrs. B.	94	1038·5
E.	93	1042
F.	92	1040
M.	88	1048
S. B.	85	1045
T.	69	1049

In the cases given there was always a slowing of the pulse-rate as improvement took place, and in obstinate cases its frequency shows little or no tendency to abatement. (See Charts IV.-XI.) If a patient who is under treatment shows a fresh acceleration of the pulse, an examination of the blood will nearly always show that there has been some relapse.

Hayem states that the pulse - rate is sometimes accelerated when the anæmia is severe; but he states definitely that the pulse-rate follows no rule. Hayem's observations on the rate of flow of blood in capillary tubes show that chlorotic blood flows through fine tubes very readily. I can confirm his observation, and it seems to me possible that in cases of chlorosis the peripheral resistance may be so much diminished as to lead to the alteration in the pulse-rate.

Another point worth attention is the frequency of a *peculiar condition affecting certain of the skeletal muscles*— a condition which Joffroy * first described as a sign which he had observed in cases of Graves' disease. On asking a healthy person to look up at the ceiling suddenly, without moving the head, the eyebrows are raised and the forehead is thrown into horizontal folds by a contraction of the anterior portion of the occipito-frontal muscle. In many chlorotic women this associated movement of the occipito-frontal muscle is wanting, although the patients can contract this muscle quite easily if they try to do so. I am not aware that any explanation of this phenomena has been brought forward. The skeletal muscles of the chlorotic girl are paler than normal, and it is a well-established fact that the irritability of such a muscle increases with the amount of myohæmoglobin; further, that the irritability to electric stimuli has been shown to be lessened in chlorotic women.

I am disposed to believe that the impassive look worn

* *Progrès Medical*, t. xviii., No. 51, December 23, 1893, p. 479.

by chlorotics—and possibly the look of languor and sadness of which Hayem speaks—is related to this lessened irritability of their facial muscles, which is perhaps associated with diminished irritability of the motor nerve-mechanisms.

Tissue Change in Chlorosis.

Authors are somewhat at variance on the subject of the activity of metabolism in chlorotics.

Bohland* and R. Meyer† found that in girls with 30 and 35 per cent. hæmoglobin on Fleischl's scale the amount of oxygen consumed and the amount of carbonic acid exhaled were greater than normal, and diminished during recovery.

Hannover‡ found that the excretion of carbonic acid was increased in chlorosis.

Herberger§ found a marked diminution of the amount of urea in a thousand parts of urine in a chlorotic girl, the amount being quadrupled after ferruginous treatment.

Moriez,|| who quotes Herberger and Hannover, found himself that the amount of urea varied, being usually, but not invariably, below the normal amount.

Hanot and Mathieu¶ found a diminution of the amount of urea, and noted a relation between the number of corpuscles and the output of urea.

Ketcher** found the quantity of urine about normal,

* K. Bohland, 'Ueber den Respiratorischen Gaswechsel bei Ver-schieden en Formen der Anämie,' *Berl. Klin. Woch.*, 1893, s. 417.

† R. Meyer, *Ueber den Saurestoff Verbrauch und die Kohlensäure Ausschiedung bei den verschiedenen Formen der Anämie.* Dissert., Bonn, 1892.

‡ Hannover, *De Quantite Acidi Carbonici*, Hanniæ, 1845.

§ Herberger, Buchner's *Repertorium*, vol. xxix., 1843, p. 236.

|| Moriez, *La Chlorose.* Paris, 1880.

¶ Hanot et Mathieu, *Arch. Gen. de Méd.*, December, 1877.

** Ketcher, *Wratsch. St. Petersburg*, No. 46, 1890.

and the specific gravity and the amounts of urea, uric acid, phosphates, chlorides, and sulphates decreased.

Lava* found a diminution in the output of urea, and an increase of the output of urea during recovery, which was greater than the increase in the amount of hæmoglobin.

So far as these data enable one to form an opinion, they point to an increase of the oxygen taken up and of the carbonic acid exhaled, but to a diminution of the amount of urea excreted.

Chlorosis and Gastro-Intestinal Disturbances.

We have abundant evidence to show that chlorotic girls are liable to gastro-intestinal congestion, to gastro-intestinal hæmorrhage, with or without ulceration, and to various disturbances of digestion.

The frequency with which gastric dilatation occurs in chlorosis has attracted the attention of many observers. Hayem found evidence of dilatation in nine out of sixteen in-patients, and in eighteen out of twenty-one out-patients.

Pick† found an atonic and frequently a dilated condition of the stomach in chlorosis, which he regarded as causal, and he treated sixteen chlorotics with good results by washing out. He believed that chlorosis is brought about by the action of the poisonous substances produced by decomposition of albuminous bodies within the dilated stomach, and which produce their effects after absorption.

Luton‡ believes that ulcer of the stomach is the cause of the majority of cases, but that intestinal and other hæmorrhages are occasionally the determining cause.

* Giovanni Lava, *Gazett. Med. di Torino*, May 31, June 14, 1894.
† A. Pick, 'Zur Therapie der Chlorose,' *Wiener Med. Wochenschrift*, 1893, s. 1663.
‡ Luton, *Soc. Méd. de Reims*, Bull. No. 10.

Stockman* has recently expressed similar views.

Meinert† has pointed out in an elaborate and able paper that chlorotic girls usually suffer from a kind of gastric dilatation, a peculiar distensibility of the organ with displacement downwards (gastroptosis of Glénard). In this condition he professes to have found the anatomical basis on which the disease depends.

He further gives facts which go far towards proving that tightness of the clothing is to blame for this condition of the stomach, and that it makes its appearance after the age at which the corset is adopted. I can corroborate his statement as to the frequency with which chlorotic girls exhibit this abnormality, but the same condition is not very uncommon in men, and in women who are not anæmic, so we must seek further for an explanation of the origin of chlorosis.

A more likely explanation, it seems to me, would be that the gastric dilatation, perhaps contributed to by pressure, is brought about by the same nervous influences which produce the gastro-intestinal congestion.

Dr. Williams of Liverpool maintains‡ that chlorosis is always the result of gastric ulcer, due to the habit of wearing stays. Meinert does not go so far as this; he does not presuppose the existence of actual ulceration in each case of chlorosis.

Garrod§ maintains that hæmoglobin is built up in the intestinal mucous membrance, that the amount in each corpuscle is larger after a meal. He gives the following facts observed by him :

* Stockman, 'The Causes and Treatment of Chlorosis,' *Brit. Med. Journ.*, December 14, 1895, p. 1473.
† *Zur Ætiologie der Chlorose.* Wiesbaden, 1894.
‡ 'Chlorosis,' *Liverpool Medico-Chirurg. Journ.*, January, 1892.
§ Archibald Garrod, 'On Hæmatoporphyrin as a Urinary Pigment in Diseases,' *Arch. of Path. and Bact.*, i. 195, 1892. Quoted by Van Noorden, *Berl. Klin. Woch.*, No. 34, *op. cit.*

In dogs at the height of digestion he found, as the mean of forty experiments, in the

	Red Blood Corpuscles.	Hæmoglobin.
Mesenteric artery ...	5,362,000 ...	78¾ per cent.
Mesenteric vein	4,540,000 ...	78¾ ,,

The single corpuscles were 18 per cent. richer in hæmoglobin after the blood had passed through the intestinal mucosa.

In fifteen men he found before meals 4,905,000 blood corpuscles, and 86 per cent. hæmoglobin ; three or four hours after, 5,044,000 blood corpuscles, and 89 per cent. hæmoglobin. Both corpuscles and hæmoglobin were increased, especially the latter.

He regards it as likely that a certain body which he has separated from the urine of chlorotics is the product of a process which goes on in the intestine of the chlorotic, and which destroys the fore-stages of the hæmoglobin molecule.

I am quite ready to believe that these facts pointed out by Garrod are of very great importance, but cannot agree with his conclusions. I am disposed to think it is far more likely that during the passage of the blood through the intestinal vessels the oldest or faded red corpuscles become destroyed, the end result being that the portal blood contains fewer red corpuscles, but these richer in hæmoglobin.

If some red corpuscles are broken down in the gastro-intestinal area, the freed hæmoglobin must pass to the liver. If Garrod's facts are right—and it appears to me that they are—in all probability hæmoglobin undergoes destruction in the gastro-intestinal area in the healthy person, and contributes to the deposit of iron (siderosis) in the normal liver.

If this be the case, it may easily be supposed that the

changes which occur in the alimentary canal in chlorotic women may be the determining cause of a further destruction of hæmoglobin.

Experiments of my own suggest that the gastro-intestinal derangements of chlorotics may be the result of disturbances of innervation. In my report to the Scientific Grants Committee two years ago, I showed that after section of the splanchnic nerves in dogs, or after division of the cord high up in the neck, the injection of saline solution—contrarily to what happens when these structures are intact—led to a more permanent lowering of the specific gravity of the blood, and that the mucous membrane of the stomach and intestine became intensely congested, of a deep chocolate hue, and copious capillary hæmorrhage occurred very readily. Immediately after death, in one case, several intensely congested patches were found in the stomach.

I had under my care an anæmic young woman (R.), who suddenly had copious hæmatemesis, which threatened to prove fatal. It appeared at the time when menstruation should have occurred. She had marked gastric dilatation, and gastroptosis, but no pain, tenderness, nor vomiting before or after the attack of hæmatemesis. She is now perfectly well, her blood is rich in hæmoglobin, her digestion perfect; and I am ready to confess that, without an examination of the blood, the case might have passed for one of chlorosis if the hæmatemesis had not been observed.

This leads me to mention the suggestion which has been made by v. Hosslin that chlorosis is due to occult gastro-intestinal hæmorrhage. In order to settle this question for ever, as it should soon be settled, we need an increased number of careful estimations of the iron in the fæces of healthy and chlorotic women, an investigation which I have not been able to undertake hitherto.

The effects on the blood of copious and repeated

CHART V.

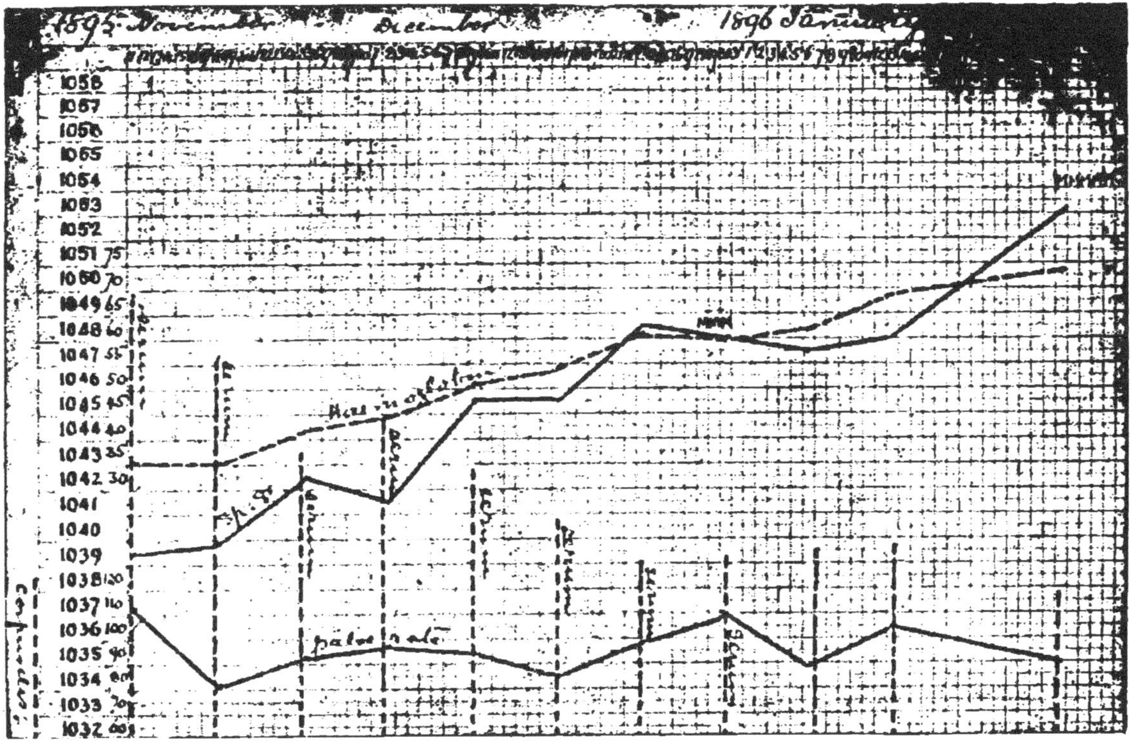

1896 January

Metallic iron, two and a half grains daily.

CHART VI.

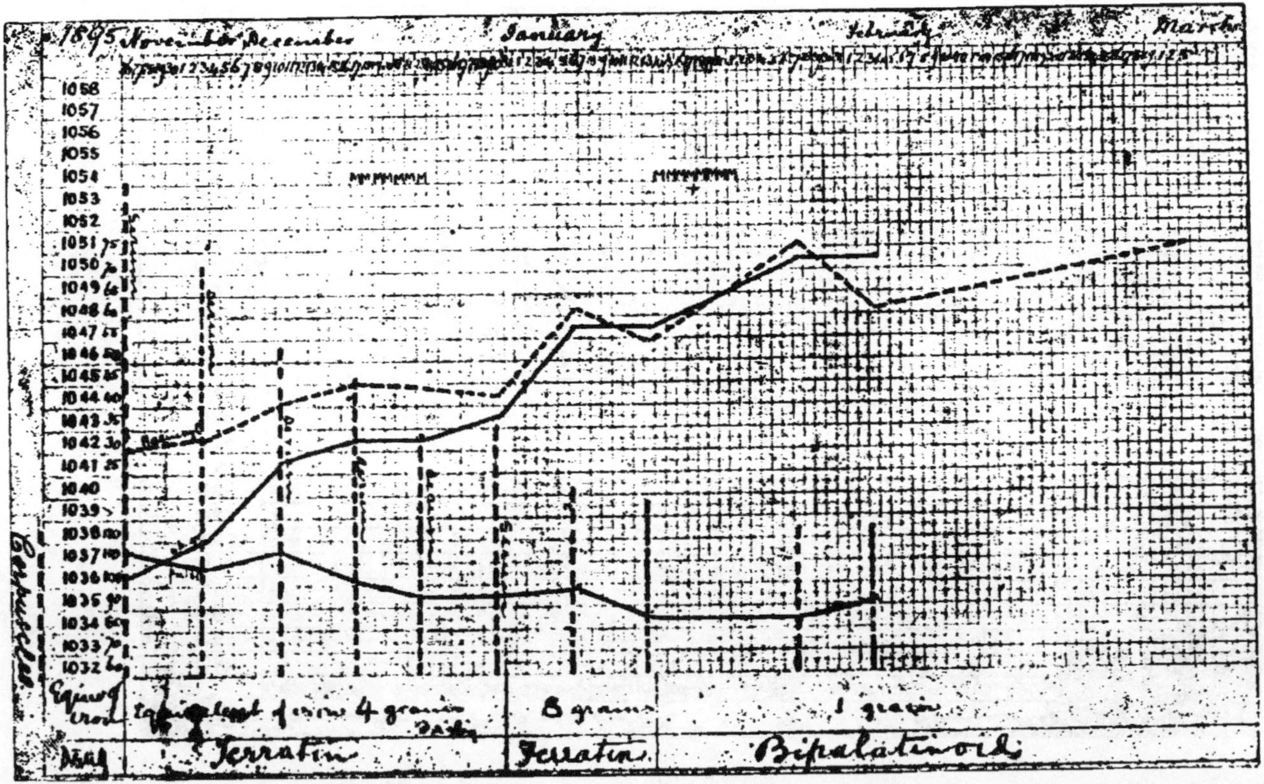
1895
November
December
January
February
March
1058
1057
1056
1055
1054
1053
1052
1051
1050
1049
1048
1047
1046
1045
1044
1043
1042
1041
1040
1039
1038
1037
1036
1035
1034
1033
1032
Corpuscles
Equiv. of Iron
8 grains
1 grain
Drug
Ferratin
Ferratin
Bipalatinoid

bleedings were investigated by Hayem.* In an en-
deavour to produce a chlorotic condition of blood, he
claimed to have produced a condition of blood analogous
to that we find in chlorosis; but he appears to have
disregarded the condition of the plasma.

All authorities agree that the plasma is practically
unaltered in cases of chlorosis, while with as great
unanimity all agree that the plasma is changed in the
anæmia of hæmorrhage. Immerman and others have
found this to be the case, and my figures show the
same.

If chlorosis were due to hæmorrhage, it would be truly
remarkable that the plasma should remain of a normal
specific gravity, or even be richer in solids than the
healthy serum. To my mind, the supporters of the
'hæmorrhagic theories' of chlorosis may be right, but
they have hitherto brought forward very little evidence in
support of their theory.

The peculiar disturbances of the appetite which are so
common among chlorotic girls are of great interest. I
have generally noticed a fondness for carbo-hydrate food-
stuffs, and especially starchy foods. Almost all chlorotic
girls are fond of biscuits, potatoes, etc., while they avoid
meat on most occasions, and when they do eat meat, they
prefer the burnt outside portion, which has, I suppose,
little value as a nitrogenous food-stuff.

Professor Stockman has recently published a series of
important observations on the amount of iron contained
in the dietaries of healthy persons. He finds that, whereas in a healthy woman the in-take of iron is about six to
eight milligrammes per day, the dietary of a chlorotic
woman contains, as a rule, much less iron. This fact
appears to me of the greatest theoretical and practical
importance. The greatest fault of the blood in chlorosis
is deficiency in iron, and yet the amount contained in the

* Hayem, *Du Sang*, p. 730.

dietary of the chlorotic is, according to Professor Stockman, about one-third of that required by a healthy woman. I insist, then, upon my patients eating rather underdone meat two or three times a day. I am disposed to regard the dietetic errors of the chlorotic as *possibly* the immediate cause of the deficiency of hæmoglobin, but we must seek further in order to account for the peculiar likes and dislikes exhibited by chlorotics in this respect.

Febrile Chlorosis.

In many cases of chlorosis the temperature will be found to be as high as 100° F.

These cases are, in my experience, those in which the signs of gastritis are dominant. The presence of fever in these cases is readily overlooked, but if the temperature were taken in all cases I believe that they would be found to be common, for I generally find at least one or two among every twenty or thirty chlorotic out-patients.

Relation to Disorders of the Sexual Organs.

My observations seem to corroborate those of others who have found that the menses are, as a rule, suppressed, or have been never properly established. Among 57 of my first 118 cases, 16 considered themselves regular, 6 had never menstruated, their ages being 16, 15, 18, 15, $18\frac{1}{2}$, and 19; in 14 the amount was diminished, in 17 there was amenorrhœa of some months' duration, and 4 had excessive loss.

Moreover, I believe with Schultze that the appearance of the menses, or greater abundance of the discharge, are always signs of amelioration. Considering the numerous investigations which have been devoted to this part of our subject, it seems that this line of inquiry has taught us less about the origin of chlorosis than might have been expected.

Rokitansky* first pointed out that anomalies of the generative apparatus and of the vessels were sometimes associated with intractable chlorosis.

Virchow† (loc. cit.) found that the generative organs were sometimes under, sometimes over, developed in chlorotics.

Fränkel‡ maintained that chlorosis and defective development of the genital organs are frequently, but not invariably, associated with defective growth of the heart and aorta; that the sexual organs exert such an influence on the body that sexual aplasia may be the primum movens of chlorosis. He held that menorrhagic forms of chlorosis are associated with excessive development as well as with aplasia of the sexual organs.

Sometimes when we watch the blood from day to day in persons who have chlorosis, one notices that about the time when they ought to be menstruating, or when they are menstruating, the blood undergoes a change akin to that which is to be noted in some healthy women about these periods.

My attention was first drawn to this point by observing in the case of G., whose blood-chart (XII.) is given, that two relapses took place with an interval of twenty-eight days. A similar change was to be observed in some other cases.

I believe that sudden relapses in chlorosis take place most frequently at these times even when menstruation does not occur. Immerman, whose able article on chlorosis in Ziemmsen's Cyclopædia is one of the best

* Rokitansky, Handb. der Path. Anat., 1846 ; Lehrb. der Path. Anat., 1856.

† Virchow, Beitr. zur Geburts. und Gynack., Berlin, 1872, t. i., p. 323.

‡ 'Ueber die Combination von Chlorose mit Aplasie der Weiblichen Genital-organe,' von Ernst Fränkel, Arch. f. Gynækol., Bd. vii., s. 465.

monographs on the disease, states that in a typical case the invasion usually occurs in connection with a particular menstrual period. Vierordt (*loc. cit.*, p. 107) found in the case of a chlorotic patient that the reappearance of the menses in abundance was accompanied by some tendency to relapse. He refers to the statement of Immerman, that although healthy women bear the menstrual loss without ill consequences, in weakly persons menstruation, acting concurrently with other influences, may produce anæmia.

In some cases chlorosis is clearly aggravated by menstrual loss, but there are many cases which cannot be explained in this way, *e.g.*, in those who have not menstruated, or who have amenorrhœa. Cases like that of R., and the cases quoted by Pidoux and others, make it appear likely that both the chlorosis in some cases, and the gastro-intestinal hæmorrhage in others, depend upon some change which takes place about the menstrual periods—a change in the production of which the nervous system has a share ; but there is not sufficient evidence to show that chlorosis is due to the menstrual loss.

So far a rather incomplete account has been given of the pathological and clinical aspects of chlorosis, and if some important matters have not been touched upon, it is partly on account of the writer's inability to add anything of material interest to the work of so many excellent observers.

The Pathogeny of Chlorosis.

At the present day I suppose that most of us feel that the pathogeny of chlorosis is in an unsatisfactory condition. The late Sir Andrew Clark maintained that the disease was due to auto-intoxication by fæcal products, but he always treated his cases with iron ; and further chlorosis does not arise in young men as the result of constipation, and only among young women do we find undoubted cases of the disease. Moreover, although constipation is common in all women, and especially in women who, from any cause, are taking less exercise, it is not more common in chlorotic women save as a result of injudicious medication. Many women presenting characteristic signs and symptoms of chlorosis do not suffer from constipation, as the notes of a large number of cases testify—cases which were especially observed from this point of view.

Virchow showed that chlorosis was sometimes associated with certain abnormalities of the vascular and generative system. Such abnormalities are not sufficiently frequent to account for so common a disease ; severe chlorosis is exceedingly common, and perfect and permanent recovery takes place in many of these with proper treatment.

Hayem[*] states that under-development of the vascular system occurs in tubercular subjects who die while undergoing the evolution of puberty ; and he states that under-development of the vessels is as common in boys as in girls.

Nevertheless, I am inclined to believe that there is some important relation between the capacity of the vascular system and the composition of the blood ; and I have shown elsewhere that section of the splanchnics by increasing the capacity of the vascular system does apparently lead to a relative increase of the plasma.

Dr. Cosgrave,[†] in a paper on the anæmia of puberty, quotes Beneke,[‡] who has shown ǂthat the annual increase in the heart and bloodvessels in girls before the age of

[*] *Du Sang*, p. 691.

[†] E. Macdowel Cosgrave, M.D., 'The Etiology and Classification of the Anæmia of Puberty,' *Brit. Med. Journ.*, March 31, 1888.

[‡] Beneke, 'Ueber das Volumen des Herzens und die Umfänge der grossen Arterien des Menschen in den verschiedenen Lebensaltern. Cassel, 1881.

puberty is 8 per cent. per annum ; whilst during the establishment of menstruation it is 80 to 100 per cent. If puberty be established in a single year, an extra growth in weight of 70 to 90 per cent. is entailed,' in addition to ordinary growth. It appears to me that if for any reason (such as widespread vascular dilatation) the vessels become unfilled at this age, it is likely that one of two results may follow, namely, (1) That the vessels which are incompletely filled may remain small ; or (2) if the vessels continue to grow as they should do, and new red corpuscles are not formed in sufficient quantity, the blood must of necessity be altered in quality.

These observations of Beneke's are quoted by Dr. Cosgrave in order to show that something like the second of these results occurs. But I have quoted Beneke's observations for another reason, in order to suggest that *arrest of development of the heart and blood-vessels in chlorotics may be the result of the anæmia, and not the cause of it.* I am not aware that this suggestion has been made, and it seems to me to offer the most intelligible explanation of the fact that ill-developed heart and vessels are sometimes, but not invariably, associated with the chlorotic condition.

Murri* maintains that circulatory disturbances thus induced by functional alteration of the vaso-motor centres are capable of determining modifications in the composition of the blood. Slight alterations in the composition of the blood may evoke, so he contends, a destruction of red corpuscles. Cold, according to Murri, is liable to aggravate chlorosis in this way, so likewise does muscular exercise.

Fränkel†. holds the view that sexual aplasia is the *primum movens,* but that menorrhagic forms occur.

* ' Pathogénie de la Chlorose : de l'Action du Froid chez les Chlorotiques,' *Sém. Med.,* 1894, p. 162.
† See p. 41.

CHART VII.

Equiv. of metallic iron.	About 2 grains.	About 5 grains.
Drug.	Ferratin.	Ferratin and iron creams.

CHART VIII

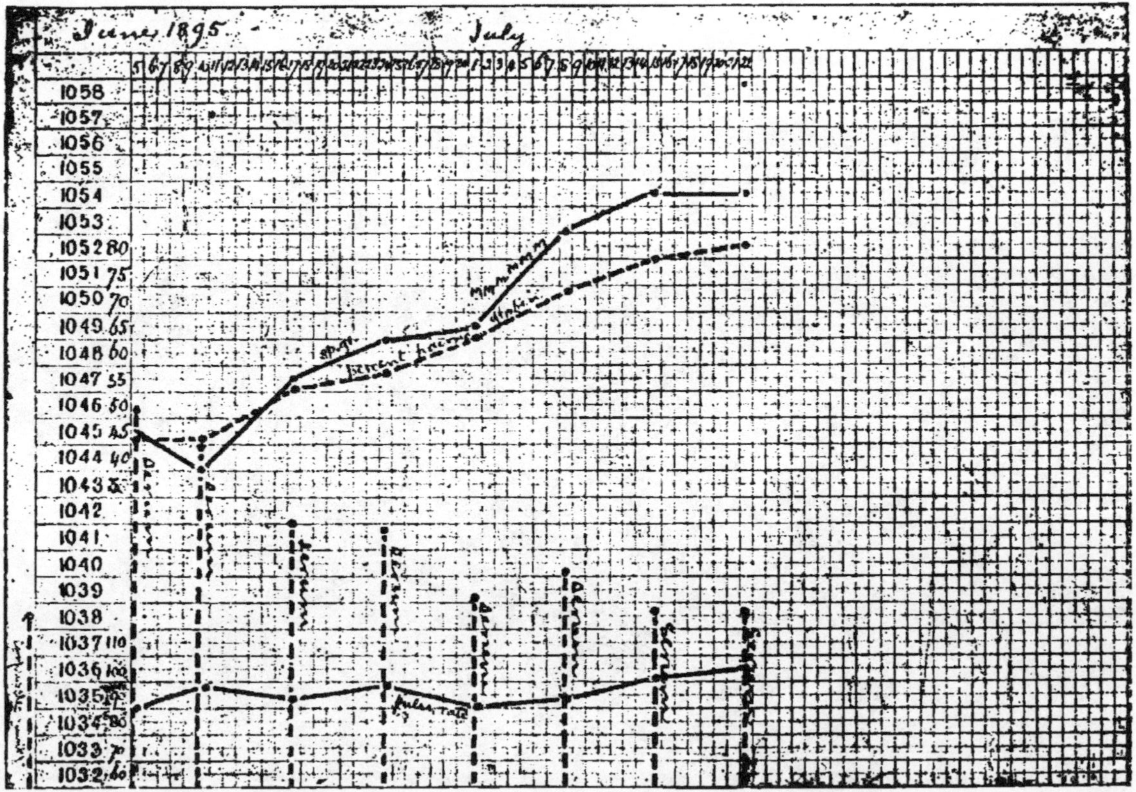

Equivalent of metallic iron, one grain daily.

Bipalatinoids three daily.

Hayem attributes it to relative excess of destruction over production of red corpuscles, due to a weakness of the hæmatopoietic tissues ; and he does not doubt the occurrence of an actually increased destruction of red corpuscles.

Trousseau, with many others, regarded chlorosis as a neurosis, and I am at one with those who hold this view so far as this, that I believe that the blood changes are produced through the medium of the vaso-motor nervous system.

Meinert has done quite as signal a service in drawing attention to the constancy of certain physical changes in the stomach which appear to him to be due to compression by the corset. But I suppose that few of us are satisfied as to the finality of any of these explanations.

Surely there is much to be said for all of the views which I have mentioned. Chlorosis has been often associated with certain affections of the generative organs (Rokitansky, Virchow, Schultze, etc.). It is associated with certain disturbances of the gastro-intestinal tract, notably gastric dilatation, ulceration, and hæmorrhage (Meinert, v. Hosslin, Luton, Williams, Stockman) ; and it is in one sense a neurosis (Trousseau).

Following in the wake of such observers, one may well hesitate to pronounce any judgment, or to suggest any new view of the disease ; but I may perhaps be excused for making some suggestions, which are only to be regarded at present as suggestions.

Seven years ago the results of observations on the specific gravity of the blood in healthy women, and on the subjects of chlorosis, led me to regard the disease as an exaggeration of a condition which, perhaps, exists in a slight degree in most young women, and which, as I have already suggested, appears to have for an end the storing up of food-materials against the event of pregnancy occur-

ring. Believing this, I concluded that in all probability
chlorotic women would have many brothers and sisters,
for the mother's blood would be likely to be similar to
that of the children, and hence she ought to have had
plenty of food-material available for the nutrition of a
fœtus in utero. Whether the reasoning was correct or
not, the results of the observations on chlorotic girls came
out as I expected.

In my report to the Scientific Grants Committee in
1893,* I drew attention to the fact that chlorotics have
usually many brothers and sisters. I explained in that
report why I looked out for this, and how my expectations
were fulfilled ; but it is since that time that I found that
in healthy persons a low specific gravity of the blood
is generally an index of great fertility on the part of the
mother, and appears to affect the children of both sexes.
In the earlier part of this paper I have written on the
subject at length, and here I will content myself with
giving the average number of brothers and sisters in some
cases of chlorosis :

Name.	Specific Gravity.	Brothers and Sisters.	Remarks.
Mrs. B. ...	1038	 16	——
T.	1042	 15	——
A. P. ...	—	 13	——
H. A. ...	1040	 12	——
E. C. ...	1042	 12	——
N. L. ...	1048	 12	——
E. W. ...	—	 12	——
M. M. ...	1038	 10	——
F. N. ...	—	... 10	——
S. K. ...	1035	 10	——
F. S. ...	—	 10	——

* 'Preliminary Report on the Causes of Chlorosis,' *British Medical
Journal*, September 23, 1892.

Name.	Specific Gravity.	Brothers and Sisters.	Remarks.
K. C. ...	1044	... 10	——
F. G. ...	1041·5	... 10	——
B.	1040·5	... 9	——
F. Y. ...	1048	 8	——
Mrs. G. ...	1043	 8	——
E. B. ...	1049	 8	——
J.	1042	 7	——
S.	1047	 6	——
M. S. ...	1038	... 6	——
I. W. ...	1041	 6	——
F. B. ...	1049	... ... 6′	——
T.	1049	 5	——
E. N. ...	1043	... 4	——
A. T. ...	1042	 4	——
— ...	—	 3	——
M. S. ...	1048	3	{ Father died a month after birth of last child.
E. J. B. ...	1036·2	 2	{ Father died when patient three years of age.
E. T. ...	1036	... 2	{ Father died when patient nine months old.
A. R. ...	1036·5	 7	——
F. C. ...	1040	 2	——
L. B. ...	1036	 1	{ Mother died in confinement.
— ...	—	... 1	——
A. M. L. ...	1037	... 0	——
J. A. W. ...	1041	... 0	{ Mother died in only confinement.

I find that a man whose blood has a specific gravity of 1060 has five to six brothers and sisters. A man whose blood has a specific gravity of 1056 has eight or nine brothers and sisters. An average healthy woman of eighteen has a specific gravity of 1054; her brothers and sisters should be, according to the works on Obstetrics, 5. A chlorotic has a specific gravity of 1030-1045, and the number of brothers and sisters is usually large, the average

4

in the above cases, excluding five of them, being eight. This shows that the blood change may be regarded as analogous to that which goes hand in hand in healthy people with fertility of the mother.

There is little doubt that chlorosis is to some degree hereditary; it affects the females of certain families, the males being, of course, unaffected. My observations, however, lead me to believe that the blood of the males belonging to chlorotic families approaches in character the blood of women.

Further observations, which were in part reported in subsequent years, led me to attach great importance to the gastro-intestinal changes, and to the part played by the nervous system in the production of chlorosis; and at Newcastle I confessed that I was almost at one with those who follow Trousseau in regarding the disease as a neurosis.

It seemed to me that chlorosis was associated with some alteration of the vaso-motor mechanisms; and I believed then that reflex, widespread vascular dilatation, most probably affecting the splanchnic area in particular, and induced in a reflex manner by centripetal impulses originating in the sexual organs, might account at once for the gastro-intestinal lesions so common in chlorotics, and for the altered condition of the blood. I assumed that, in some way or other, alterations in the blood were the result of this widespread gastro-intestinal congestion.

My own experiments above cited showed that section of the splanchnic nerves might induce blood changes. Garrod's facts now appear to me of much importance in this connection, although I cannot accept his interpretation of them. He concluded from his experiments that hæmoglobin was built up in the gastro-intestinal vascular area; these experiments seem to me to point to the fact that even in healthy persons some red corpuscles are broken down in the gastro-intestinal vascular area, and one can well imagine that this process of hæmolysis is increased in chlorosis. If this be the case, it may account for the fact that v. Hosslin found an increased amount of iron in the fæces of chlorotic girls. Hayem, whose opinion must have great weight, speaks of an increase in urinary chromogen, and of increased hæmolysis. At the same time observations of G. Hoppe Seyler and of Garrod point to an actual diminution of the urinary and fæcal pigments which result from the decomposition of hæmoglobin. I cannot reconcile these contradictory observations, though I notice invariably that the urine of chlorotics contains a chromogen which reduces Fehling's solution. The question cannot be decided at present; but if Garrod's observations are correct, some destruction of red corpuscles must, I think, go on in the gastro-intestinal area of the healthy individual; and one can well believe that the gastro-intestinal disturbances, which are so constant in chlorotic women, are accompanied by an exaggeration of this process. Such changes of this kind as occur in the healthy woman might, when within physiological limits, be of service to the organism by leading to a diminution of tissue-change, a conservation of nutritive materials; and this view does not seem to me unreasonable, in spite of the fact that tissue-change is certainly not always, possibly never, lessened in chlorotic persons.

Some years ago I was struck by the analogies between Graves' disease and chlorosis—two diseases which are regarded by nearly all authorities as intimately related. The first promulgation of the now accepted doctrine that Graves' disease is a thyro-intoxication led me to ask myself whether chlorosis might not also be due to a somewhat analogous process. And my suspicions at first rested on the thyroid. The clinical picture in the case of each of these two diseases is very similar in some respects.

In both conditions we have some enlargement of the thyroid and a peculiar rapidity of the pulse; in both

there is often a fine tremor of the fingers, there is the
weakening ' paralasia ' of some of the skeletal muscles to
which reference has already been made. The cases of
these two diseases, which have a sudden onset, seemed
to me to point to some similarity in their origin. For
instance, Trousseau relates an interesting case in which
Graves' disease came on suddenly in a woman who was
grieved by the death of her husband ; and Pidoux, to take
another example, relates the case of a young girl who,
while menstruating, became chlorotic after a sudden chill
caused by immersing her arms in very cold water when
she was overheated and tired. Hayem refers to several
cases of similar origin, and Professor Clifford Allbutt has
seen several such cases, one of which he related to me
privately.

Seeking for some source whence this auto-intoxication
might arise, one would naturally ask whether Sir Andrew
Clark was correct in his view that chlorosis was due to a
faecal poisoning. But it seems that both the use of purges
and the production of some partial intestinal asepsis—for
we can only bring about a partial intestinal asepsis—have
failed to effect the cure or to prevent the occurrence of
chlorosis ; while the observations of Rethers and Mörner
show definitely that the ethereal sulphates of the urine are
not increased in chlorosis, as they would be if intestinal
decomposition were more active. We must look, then,
elsewhere for some body or bodies which can produce
such characteristic and peculiar effects.

Failing to find any records of a case of undoubted and
marked chlorosis in a man, regarding the fact that the
blood of the woman only is liable to this marked alteration, bearing in mind, further, that the blood of the
woman acquires distinctive characters after the age of
puberty, I asked myself whether chlorosis might be
caused by some product or products originating in the
organs peculiar to the female, which, by acting on the

Chart IX.

Bipalatinoids.

Ferratin.

nervous system, particularly on the vaso-motor system, produce the blood changes which occur in the chlorotic woman at puberty, and in many women about each menstrual period.

It seemed to me possible that at each ovulation an absorption might take place of the products of the epithelium of the Graafian follicle, of liquor folliculi, etc.

It seemed reasonable to accuse the sexual organs first of all, because the disease always occurs at or after puberty, when a stream of unwonted substances must be poured into the blood, if not for the first time, then at least in greatly-increased quantities. The liability to chlorosis diminishes as tolerance to these is gradually acquired, or as their production is lessened.

One fact appeared to me to lend some countenance to such a suggestion, namely, that chlorotic women come of prolific mothers, and healthy persons whose blood approaches the limit between chlorosis and the average in health are, I believe, generally prolific.

Now, it might be conceived that this fertility is in part due to greater ovarian activity; but whether this be the case or not, I do not know. If the ovaries produced a product which could bring about such effects, then we might expect the blood to become altered, as it is altered at the age of puberty, when the ovary becomes more active; and, further, if ovarian products could produce such effects, then we might look for a change in the blood after the menopause, and for a similar change in those women whose ovaries had been removed by a surgical operation.

We do find, as I have shown to you, that the blood of the female nearly resembles that of the male, both before puberty and after the menopause.

But I turned with some anxiety to an examination of the blood of some patients upon whom oöphorectomy had been performed. By the kindness of Dr. W. S. A.

Griffith, Dr. Joseph Griffiths of Cambridge, Dr. Eden, and Mr. Christopher Martin of Birmingham, I was allowed to examine the blood of seventeen of these women, and I was unable to ascertain that the removal of the ovaries had produced any constant alteration in the blood so far as the specific gravity or the amount of hæmoglobin were concerned. I hope to discuss the subject more fully in a subsequent paper.

I am at the present time endeavouring to ascertain which organs may produce the toxic bodies to the effects of which I am inclined to attribute chlorosis, and I am seeking especially to ascertain the influence of *uterine* products.

With regard to the supposed occurrence of chlorosis in males, it appears to me that if the same disease could occur in males, surely there would have been found one case at least presenting the same signs which attend it when it attacks females. In chlorosis the specific gravity of the whole blood commonly falls twenty degrees, but I have not been able to find a single case of anæmia in a male subject, free from serious organic disease, whose blood specific gravity has fallen twenty degrees, nor ten degrees.

The male blood is liable, as the result of serious organic disease, to profound alterations; males, for instance, with heart disease may suffer from an intense anæmia; they may wear the aspect of chlorotics, their blood may in some respects be indistinguishable from that of a typical chlorotic girl. But exclude those who have heart disease and renal disease, or other organic disease, and I find no anæmia in males which is comparable in severity with the anæmia of young women. Among many hundreds of young men, I have never detected one who had only 17 per cent. of hæmoglobin, or a blood specific gravity of even 1045, while in the same number of young women this would certainly be the case. The only explanation appears to me to lie in the supposition that chlorosis does not occur in the male.

'My observations appear to me to indicate that chlorosis is an exaggeration of a physiological blood condition, an exaggeration of a change which occurs in the blood of healthy females at puberty, and which shows itself in many females at each menstrual period. This physiological change has a conservative tendency, so far as metabolism is concerned. It tends to lessen metabolism, so that food-materials are saved up against the event of pregnancy occurring. It is a preparation for motherhood.

Chlorosis is related to fertility on the part of the mother, and possibly to facultative fertility on the part of the children.

Further than this, I am only in a position to suggest that appearances point to the likelihood of chlorosis being a chronic auto-intoxication brought about by some substances which are possibly the products of uterine, Fallopian, or ovarian metabolism, and producing their effects on the blood by inducing changes in the gastro-intestinal canal by the medium of the nervous system. The wearing of corsets materially increases the liability to chlorosis, and if my views of the significance of the disease are well founded, then we have a *raison d'être* for the adoption of this article of dress.

PART III.

THE PREVENTION AND TREATMENT OF CHLOROSIS.

BEFORE discussing the therapeutic measures which may be adopted in order to cure chlorosis, it is well that we should refer to the means which may be taken to prevent its development.

It seems to me that we have two potent prophylactic agents—proper exercise and proper diet.

Among young women who take plenty of open-air exercise, I find that the blood is richer in hæmoglobin than among other young women; it approximates the male blood in characters. It is certain that want of exercise more than anything else is in many cases a powerful determining cause of chlorosis.

Healthy young girls and young women ought to take open-air exercise as young men do, and, fortunately, it is becoming more customary for them to do so.

I am constantly advising women to use the bicycle, for it has been stated that it tends to lessen menorrhagia, and I have found some severe cases which yielded at once to this treatment alone. It may be that it tends to increase the circulation through the muscles, and to lessen the activity of the sexual organs, and if so, it should be found beneficial in those predisposed to chlorosis, if my views of the etiology of the disease are correct.

It is of no less importance that growing girls, and

especially those who may be suspected of a predisposition to chlorosis, should be made to eat proper food in proper amount.   Iron appears to be taken up from the iron contained in organic compounds, such as the muscle fibre of beef and mutton, and the chlorophyll of green vegetables, and these food-materials are necessary, that the blood of young women may be furnished with a due amount of hæmoglobin.   Anæmic young women usually eat little or no meat, and Professor Stockman's experiments confirm the view that chlorosis is largely due to a dietary containing too small an amount of iron.   The depraved appetite which leads these young women to choose such an unsuitable dietary is, I take it, a result of the morbid condition which gives rise to chlorosis.   It appears possible that a morbid dislike for all manner of iron-holding foods may be one means by which the blood-change is brought about.

The above measures of prophylaxis are, I think, entirely sufficient to prevent the onset of chlorosis in most girls, and yet, despite these precautions, chlorosis will sometimes occur.   I have known chlorosis develop in the case of a young farm-servant, who was working every day in the hayfields, and who had every opportunity of eating suitable food ; but I much doubt whether she ate sufficiently of those food articles from which iron can be assimilated.   In many cases the patient asserts that she does eat meat ; but on inquiry, it is found that she always chooses the burnt, over-cooked parts of a joint, and leaves the rest.

The Hygienic and Therapeutic Treatment of Chlorosis.

In considering the therapy of the disease, for convenience I sub-divide cases of chlorosis into three categories : (1) Simple chlorosis ; (2) Chlorosis with gastritis, or with

gastric ulcer; (3) Chloro-oligæmia. Akin to these last
are cases of simple oligæmia, which have arisen indepen-
dently of chlorosis, a form of anæmia which occurs also in
men, usually in those belonging to small families, and with
which I do not intend to deal in this paper.

1. In cases of *simple chlorosis*, however severe, I do not
require the patient to adhere to any fixed rules with regard
to diet, etc., beyond pointing out the advisability of keep-
ing good hours, of eating underdone meat twice a day,
and partaking of green vegetables, and of taking some
exercise in the open air daily.

I give these patients a few bipalatinoids or a little
reduced iron three times a day. I scarcely ever order
them purgatives. At one time I gave an aloetic pill night
and morning, but I am convinced that, as a rule, the constipation rarely calls for treatment, and in time the iron
cures their constipation as well as their chlorosis, provided
that an overdose or an unsuitable preparation be not given.
For the headaches, which are so prominent a feature of
chlorosis, I am in the habit of prescribing very small
doses of salicylate of soda and of antipyrin at short
intervals during the attack, telling the patient to avoid
taking the medicine as much as possible, and assuring her
that, when her blood state is improved, the headaches will
cease to give trouble.

2. *Chlorosis with gastritis.* — In these cases there is
usually pain after food, especially if meat be taken ; there
may be vomiting, and there is often an elevated temperature.

In these cases I enjoin rest, and tell the patient to take
two pints of milk with or without soda-water during the
twenty-four hours. After a day or two she may take a
little underdone meat, at first once, and then two or three
times a day, taking no potatoes, very little bread, and no
stimulants. I order bismuth salicylate (with morphine in
some cases) to be taken before food, and a bipalatinoid

directly after food three times a day. The bismuth is continued until the pain ceases, or until the temperature is normal. The subsequent treatment is that for simple chlorosis.

If there be distinct signs of gastric ulcer, or signs pointing to the probability of ulceration, these cases need to be kept in bed. In some of them feeding by the mouth is inadmissible, and enemata of peptonized milk and peptonate of iron may be used, giving bismuth, or bismuth with morphine, by the mouth. In cases attended by hæmatemesis, ice, or iced hazeline, may be used internally, and an ice-bag may be applied to the epigastrium.

After a few days peptonized milk, or milk and soda, may be given by the mouth, continuing the enemata of iron, or giving a few bipalatinoids daily. As soon as possible, raw meat-juice should be added to the dietary, but the patient must be told to avoid eating much bread or potatoes, and to take chopped spinach or other green vegetables.

3. *Chloro-oligæmia.*—These cases, when severe, are best treated by rest in bed, with generous diet, and maltine and cod-liver oil with iron. After a time they may be allowed to commence taking exercise daily, and a prolonged holiday at the seaside is often of the greatest service. But in all cases of chloro-oligæmia treatment of every kind is somewhat unsatisfactory, and I cannot say whether these patients ever get really well if the disease is of long standing.

Too much stress cannot be laid on the importance of examining the blood in every case of chlorosis before it is pronounced 'cured,' for the appearance of the face is not infrequently misleading, even to a practised eye, and the percentage of hæmoglobin may be very low in a girl who has a brightly-coloured complexion. I never regard my cases as cured until the number of corpuscles, the amount of hæmoglobin, and the specific gravity of the blood are

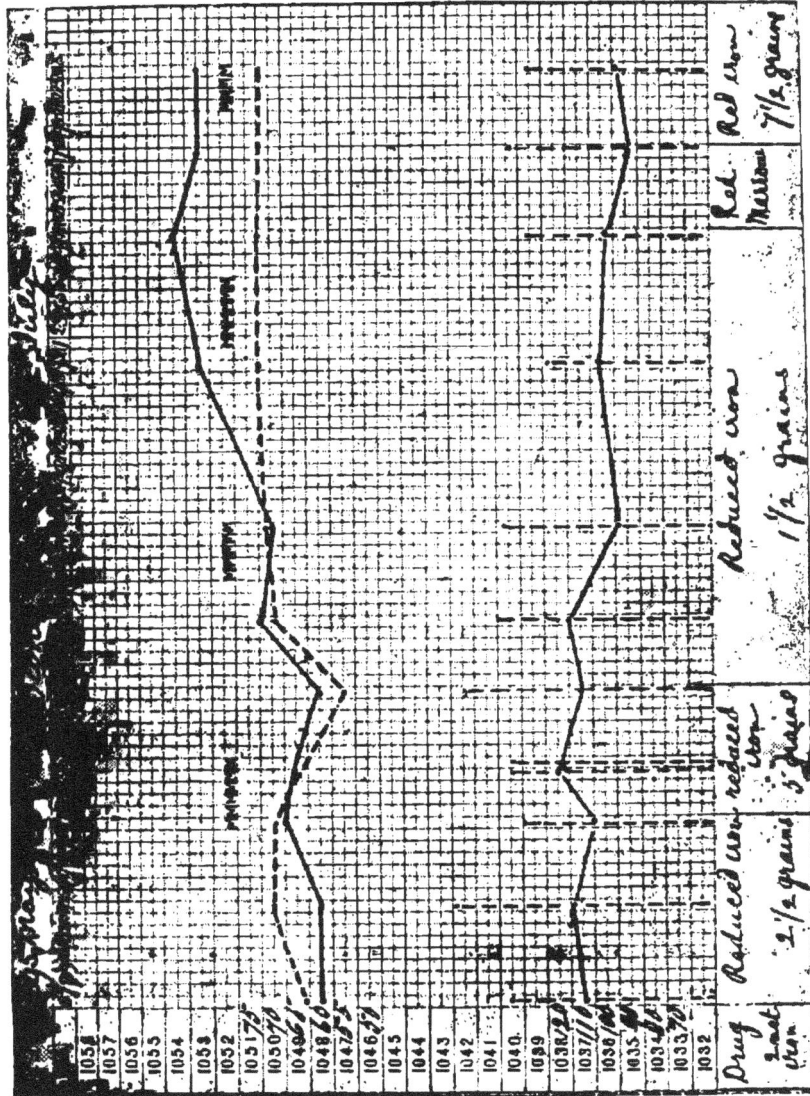

CHART XI.

CHART XII.

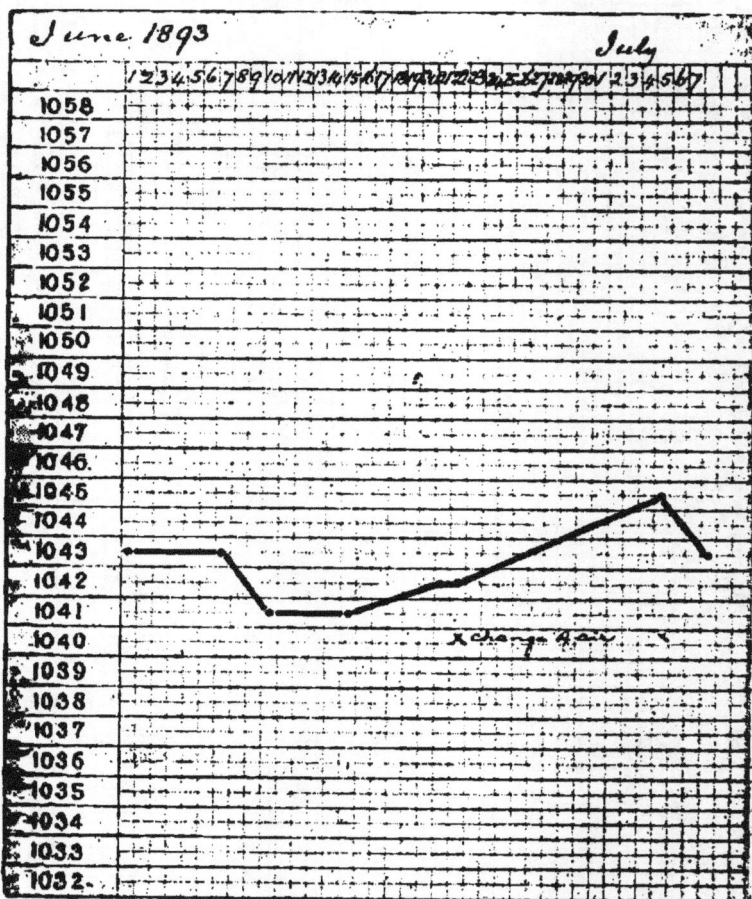

quite normal. The return of the healthy state takes place quickly in some cases—I have known this happen in a week—and in other cases may not take place for months, or for more than a year.

The condition of the blood before the treatment is commenced affords no indication of the rapidity with which the disease will yield to treatment. Some of the cases in which the percentage of corpuscles and hæmoglobin are the lowest are the easiest to cure ; while the intractable cases are, I think, most often those which I have termed ' chloro-oligæmia ' (see Charts XI., XII.), in which the percentage of hæmoglobin and the proportion of corpuscles are not so far removed from the healthy limits. I may be allowed to repeat that these latter cases are most frequently of long duration, and I am wont to regard them as due to a deficiency of plasma, which may indeed mask an actual deficiency of hæmoglobin, or of red corpuscles.

I have already expressed my doubt whether chloro-oligæmia can be cured, and it is certainly difficult to cure when of long standing. It follows an antecedent chlorosis which might have been cured, but which has been neglected or but partially cured. It is, therefore, of the utmost importance that chlorosis should be treated at its onset, especially in girls at about the age of puberty, lest, neglected, it should lead to a chronic state of ill-health.

The Treatment of Chlorosis with Iron.

Iron.—Iron has long been recognised as a specific for chlorosis, but even at the present day opinions differ as to the best mode of administering it.

I have made a large series of observations on chlorotic patients with the object of deciding, if possible, what form of iron gives the most rapid and satisfactory results (Charts IV.-X. and XII.).

Some cases yield at once to the administration of almost any preparation of iron, others prove almost equally intractable to all of them.

The results of some of my observations are shown in Charts IV.-XII., which are selected from some forty somewhat similar ones. In each case general notes were taken, and once a week the blood was examined, a note being made of the specific gravity, the percentage of hæmoglobin, and the proportion of serum to corpuscles.

On each chart I have represented the results, together with the pulse-rate and the dates of menstruation on each occasion. The curves showing the percentage of hæmoglobin, the specific gravity of the blood, and the pulse-rate hardly call for explanation ; but a word is necessary to explain that the vertical lines show the proportion of serum to corpuscles, the column of corpuscles being reckoned as unity. The unit of length of the corpuscular column adopted is from the base-line to the line marked 1038. Now, as a chlorotic patient gets well, one invariably finds that the lines indicating variations of the specific gravity and the percentage of hæmoglobin rise, while the lines representing the variations of the pulse-rate falls, and the proportion of serum diminishes. At the same time menses reappear if they had ceased, and the patient assumes a healthy and generally a markedly coloured complexion. If these changes take place quickly, as in Charts IV.-IX., the patient is doing well ; if the improvement be slow, or if there be any relapse, attention is at once drawn to it by a glance at the chart.

The *amount of iron* necessary to cure a case of chlorosis is generally a daily dose of about two grains. Neither in the form of sulphate, nor as reduced iron, nor as ferratin, is one grain of metallic iron so efficient as one grain of the metal in the form of the ' bipalatinoid ' of Blaud's pill.

The time of administration of the iron has an important influence on the results obtained by its use, for iron ought always to be given directly after a meal.

In comparing the effects of different preparations of iron, one has to bear in mind that cases of chlorosis present differences in the ease with which they yield to the same preparation; and one has further to take into account the amount of metallic iron which is contained in the daily dose of the preparation used in each case. Thus, ferratin contains about 7 per cent. of iron; a dose of one gramme will therefore contain about one grain of iron, and the effect of this may be compared with an equivalent dose of the metal in the form of ferrum redactum, or contained in three bipalatinoids.

In the charts subjoined the equivalent of metallic iron is always given, and it will be seen that, given in certain forms (*e.g.*, bipalatinoids), chlorosis was cured by the administration of the equivalent of a relatively small dose of iron, while given in other forms (*e.g.*, ferratin) the dose needed to contain a larger equivalent of iron before it acted so well.

I find that chlorosis may be cured with almost any preparation of iron, provided that the daily dose contains a sufficiently large amount of the metal.

I have stated that my best results were obtained with (1) ferrum redactum, (2) carbonate of iron, and I am inclined to think that these preparations probably act best because the metal contained in them easily enters into fresh combinations in the stomach. The results obtained when a very small dose of the *carbonate* was given as Blaud's pill or as a bipalatinoid thrice daily (a dose only containing one grain of the metal) were among the best, if not quite the best, I have noted, and I can only explain this by supposing that the *freshly liberated iron* is more easily taken up into some new (proteid) combination;

5

whereas the more stable preparations of iron do not so readily combine with other substances in the stomach.

Reduced iron acted well in many cases (see Chart V.).

The ammonio-citrate gave some good results, but I found it advisable to exceed the pharmacopœial limit.

Red marrow (see Chart XI.) appeared to exercise no stimulating effect on the production of hæmoglobin in cases of chlorosis. In order to do good in chlorosis it would have to be given in enormous doses, containing a sufficiently large amount of iron, and such doses might be poisonous.

Ferratin (see Charts VI., VII., IX., and X.) seems to me to offer no advantage over other preparations of iron, and the doses recommended appear to me to be much too small. At the same time, ferratin gives good results at times, though no better results than can be obtained with simpler and less expensive compounds.

In an interesting paper 'On the Effect of Ferratin in the Treatment of Anæmia,' by Drs. Jaquet and Kündig, the authors give in tabular form the results of observations made on twenty-five patients of both sexes and of all ages.

Some of these were cases of chlorosis, and in certain of these ferratin acted as well as any preparation of iron might be expected to do, but in several cases the result was not good. One of these (Agathe Bannwarth) affords a good example of an intractable case of chlorosis : the patient was twenty-six years of age, improved for a time, but a relapse took place while under treatment with ferratin, after she had been discharged from the hospital.

The case of Elise Sauer is precisely analogous in some respects. In her case no preparation of iron gave a successful result, and the improvement while taking ferratin was not marked ; the hæmoglobin in her case was 42·93 per cent., on Fleischl's scale, on January 27, 1894, and on March 10, 1894, it was only 47·4.

In treating chlorosis with iron I fear that we do not necessarily strike a blow at the cause of the disease. If my view of the ætiology of chlorosis is well founded, cases like that of Elise Sauer point to a condition of auto-intoxication, which at present we can only treat with temporary and limited success.

EXPLANATORY NOTES ON CHARTS IV.-XII.

CHART IV.

S. K. T., aged 20¾ years, simple chlorosis. Pale over three years, has only menstruated once in four years. The chart shows a great reduction of the volume of corpuscles compared with serum when first seen, and that the case yielded to very simple treatment, namely, the administration of two bipalatinoids three times a day.

CHART V.

N. G., aged 16 years, simple chlorosis. Pale four months, onset gradual, pain after food four months, never sick, bowels regular every day, not menstruated for four months, grown pale since then. She recovered speedily while taking two and a half grains of reduced iron three times a day.

CHART VI.

S. B., aged 22⅓ years, simple chlorosis. Been pale six months, menstruates regularly, and too freely. Improved with ferratin, which was increased to larger doses than those recommended : the cure was completed with bipalatinoids.

CHART VII.

M. M., aged 18 years, simple chlorosis. Not regular for two years. Improved with ferratin in daily doses containing about two grains of the metal. She recovered perfectly while taking the same amount of ferratin, with the addition of reduced iron, in the form of chocolate creams, which were prepared for me by Mr. Addison, chemist, of Cambridge.

CHART VIII.

S. B. W., aged about 20 years, simple chlorosis. Began to get pale five months ago, menstrual flow less, pain after food. Speedily recovered while taking one grain of iron daily in the form of bipalatinoids.

CHART IX.

A. C., aged 16¾ years, simple chlorosis. Pale six months, amenorrhœa four months. Did not improve much while daily taking an amount of ferratin containing three grains of the metal, but was soon cured by taking one grain of the metal daily in the form of bipalatinoids. The chart is not completed.

CHART X.

W., aged 23, simple chlorosis. Sense of fulness after food, no pain, no sickness, menstrual loss less. Very slight improvement while taking sufficient ferratin to contain three grains of the metal ; a daily dose of one grain of iron in the form of bipalatinoids cured her.

CHART XI.

E. B., aged 18½, chloro-oligæmia. Always been pale, now very pale, but lips red, menorrhagia, bowels always confined. The blood-changes were not marked, as we find in these cases, and her pallor yielded but little to treatment.

CHART XII.

Mrs. G., aged about 20 years, a pronounced case of chloro-oligæmia. No children. The case yielded to no ferruginous medication, but she improved while away for a holiday at the seaside. In Cases XI. and XII. the proportion of serum to corpuscles was near the healthy standard.

Baillière, Tindall & Cox, 20 and 21, King William Street, Strand.